Elon Musk

A Biography of the Billionaire Behind SpaceX, Tesla & Twitter

Bailey Goodwin

©Copyright 2022 by Cascade Publishing

All rights reserved.

It is not legal to reproduce, duplicate, or transmit any part of this document in either electronic means or in printed format. Recording of this publication is strictly prohibited.

Table of Contents

Introduction 1

Chapter 1: The Early Life of a Bookworm 6

    Elon's Family Life 7

    Early Education 10

    Starting Life Anew 12

Chapter 2: From Nerd to Billionaire 14

    Zip2 15

    X.com and PayPal 17

    The Price of a Vacation 18

Chapter 3: Expanding Horizons 20

    SpaceX and Starlink 21

    Tesla 24

    Neuralink 27

    The Boring Company 27

    Just the Beginning 29

Chapter 4: The Man Behind the Fame 30

    The Personality of a Genius 31

    Managerial Characteristics 32

    The Future of Humanity 35

Chapter 5: Finding Balance 37

    Relationships and Children 38

    Philanthropy Work 41

Vision on Education 43

What Spare Time? 44

Chapter 6: Twitter Wars **46**

The Cave Rescue Tweets 47

Thoughts on the COVID-19 Pandemic 48

A Dangerous Twitter Poll 50

Bill Gates Versus Elon Musk 52

Taxation Troubles 53

Other Controversies 54

Television and Film Cameos 55

Chapter 7: The Elon of Today **57**

The Boring Company 58

Neuralink 59

Tesla 60

SpaceX 62

Twitter 63

Chapter 8: To the Moon and Beyond **67**

Colonizing Space 68

Flying Electric Airplanes 69

Driving Autonomous Cars 70

Creating Ethical AI 72

Conclusion **73**

Introduction

Failure is an option here. If things are not failing,
you are not innovating enough.

—Elon Musk

May 5, 2021. Beneath cloudy skies over SpaceX's Texas facilities, history is being made. The countdown has begun. The rockets fire up. Slowly rising, Starship SN15 is lifting off, soaring into the sky. Reaching an altitude of more than 6 miles, the minimalist ship responded well, eventually returning to its landing pad. As it neared its base, specific engines reignited so that the ship would be able to execute the landing maneuver required to settle back down on its base. It's a touch and go moment, where so many other prototype rockets have failed. This time, though, the landing flip maneuver repositions the rocket vertically with success. Starship SN15 has returned safely home, changing the rocketry and space exploration landscape forever.

Starship SN15 and its record landing exist thanks to a singular question: *Why can't we have reusable space rockets?* How could we make it happen? All it took was a unique mind who wasn't afraid to question conventional thinking. All it took was an Elon Musk.

A self-made billionaire who is considered by Forbes Magazine to be the richest man on Earth, Elon Musk is estimated to own a fortune of $219 billion. His wealth, however, isn't solely what defines the man. Rather, his straight-talking, unconventional approaches, and ambitions for the planet, have pushed him to the forefront of cultural awareness. As someone who questions convention, who isn't afraid to muse aloud, and who embraces failure, Elon's life has caused waves in both the business and scientific communities. Yet, at the end of the day, Starship SN15 and the now household name of Tesla might not have existed without Elon Musk.

When I say Elon dominates the media, not only am I referring to the coverage that news outlets publish about him and his endeavors, but the fact that he broadcasts them himself. Known to have Twitter-friendly fingers and over 80 million followers on the social media platform, the billionaire isn't afraid to get on social media and communicate directly with his followers, business competitors, and enemies. With memes, emojis, and dry quips, Elon has shared his thoughts on a plethora of subjects, ranging from SpaceX liftoffs to COVID-19 to the current POTUS. As a result, Elon's bipartisan ideas and speech have not endeared him to the Establishment.

Still, for most of his audience online, Elon's quips and memes are thought-provoking and entertaining. Those who take the time to get to know Elon outside of the mainstream media coverage discover that Elon's social media profile obscures a hard-headed entrepreneur, an intelligent engineer, and a good-hearted philanthropist. As an "ideas man," the billionaire has become particularly famous for his involvement with alternative approaches to cars. The creation of the Tesla, an electrically powered motor vehicle, was part of an effort to reduce pollution and carbon dioxide emissions, and to encourage the development of more sustainable energy sources. Other projects have shown Elon's dedication to innovative thinking as a way to solve everyday problems, like resolving traffic issues or providing fast, affordable internet. At the same time, Elon has also plunged into the expensive and difficult space industry, hoping to push forward where NASA faltered. It is no surprise that, for people who place a premium on action over words, Elon's life is a model of success, power, and creativity.

An inspiration for young people who want to change the world, Elon's life provides a model for those who are willing to dream big. Entrepreneurs following in Elon's footsteps have to be prepared to take risks and make sacrifices. When we trace Elon's life story, it is clear that, while Elon's early life came from a place of privilege, the future billionaire wasn't afraid to leave his easy life behind to make his own way in the world of engineering and finance. The rollercoaster Elon experienced, one of hard work and debt eventually translating to multiple university degrees and a million-dollar business deal, exemplifies the American dream. Elon didn't stop there, either.

When many thought he would stop and settle down, Elon persisted, investing in other projects and business ventures and working his way to the top. His drive to learn, to ask questions, and to improve the world around him propelled him forward. Any discussion about Elon usually involves pointing out an indisputable fact—if Elon sees a problem, he will charge ahead and develop a solution. Elon will not sit around and wait for other people to change things for him. Instead, he gets involved. Whether he is building a company, overseeing an engineering project, or developing a program or business solution, Elon Musk enjoys taking matters into his own hands. This has resulted in Elon driving himself constantly, sleeping in his office, and refusing to relax until a challenge is overcome. Obsessed with the truth, Elon prizes empirical research and data, as well as unbiased perspectives, particularly when pursuing a solution to a business or engineering problem he is attempting to resolve. All of this points to a relentless drive to understand the world and what makes it tick.

And, what is the outcome? It's undisputable success. Elon's hard work and application to everyday problems has resulted in many surprising success stories such as the creation of Paypal and Tesla, the liftoff and landing of the SpaceX Starship, and the establishment of an alternative schooling system: Ad Astra. By the age of 50, he has achieved more than many combined, marking him as one of the most influential entrepreneurs and engineers in this century. Like Bill Gates (the man behind Microsoft) and Steve Jobs (the former founder and CEO of Apple), the Tesla entrepreneur stands as a man of vision and innovation. His projections and speculations might cause some eyebrows to rise, but Elon remains dedicated to pursuing new ways to solve ongoing concerns around the world.

At the same time, Elon has maintained a busy personal life, as well, under the scrutiny of the Press. After navigating three marriages and multiple relationships with celebrities, Elon has emerged with a large sprawling family of seven children. On top of that, he maintains firm connections with his siblings and mother. Elon often plays tennis with his brother, produced a movie for his sister, and provided for his mother financially since his meteoric rise to success. When he has free time, Elon Musk usually ends up spending time with his kids, playing video games, or reading on his inseparable iPad. When Joe Rogan asked Elon during an interview how he manages to make time for everything, Elon simply shrugged and offered a deprecating smile.

It does beg the question: *If this man has managed to achieve so much, what else can he be expected to do? Will he colonize Mars?* Sounds like a plan to Elon. *Will he get rid of fuel-based cars?* That sounds like a good idea, as well. *Should he build a Gigafactory powered by a thousand superchargers?* If it needs to be done, it will be done. *Will he restore memories and improve health with implanted brain chips?* Why not? Whatever Elon plans to pursue, you can safely assume that he won't be afraid to fail a little… or a lot. Whether he is selling a flamethrower or tequila, Elon Musk isn't afraid to take risks. Some might view these products as failures, but failure is something that the Tesla and SpaceX CEO welcomes. According to him, if there is no attempt made, there is no failure and no opportunity to learn and grow.

For those who dare to dream, Elon is someone to admire and aspire to emulate. He is a man thinking about the future, trying to transform the world into a better place. A defender of freedom and equality, Elon champions our ability, as the human race, to improve and innovate. As he once stated, "I think it is possible for ordinary people to choose to be extraordinary". To make an omelet, however, you have to break a few eggs. To those who prefer social conventions and traditions, Elon might be exasperating or dangerously reckless; but, would the world be as interesting without him? I think not. After all, who doesn't love the drama of a hilariously-timed poop emoji?

Chapter 1:

The Early Life of a Bookworm

*I was just absolutely obsessed with truth,
just obsessed with truth.*

–Elon Musk

Elon Reeve Musk, the world's richest man, is the oldest of the three children of Errol Musk, a South African electromechanical engineer, pilot, and sailor, and Maye Musk, a Canadian model and nutritionist. He was born on June 28, 1971, in South Africa, in the city of Pretoria, one of the country's three capitals. He has two siblings from his parent's union: Kimbal Musk and Tosca Musk, both also high-achieving individuals in their own fields. Elon is known to have a good relationship with his mother and siblings, but his relationship with his father has been increasingly strained over the years.

Thanks to extensive media coverage, interviews, and Ashlee Vance's biography, we have a fairly good idea about what Elon's life was like. According to him, he was a nerdy, introverted boy who was bullied in school. To overcome the loneliness of childhood, he developed an early interest in computers with the help of a machine that his father bought him. He started coding when he was ten years of age and, upon turning 12, he developed and sold the source code to his first game. The game was titled Blastar, a cult-classic space-shooter.

Elon's Family Life

While living in South Africa, the Musk family was neither wealthy nor poor; they were considered high-middle class. The information on the size of the family's wealth varies depending on where the data is obtained. According to some accounts, Elon's family had a lot of money due to his father's business endeavors. His parents remained married from 1970 until 1980, when they divorced. Elon, ten years old at that time, decided to live with his father Errol because he claimed that it was sad that his father would live alone while his mother would be with all three kids. The decision to live with him in the suburbs of Johannesburg was one that he soon came to regret. In various interviews, Elon has asserted that his father was not a great parent, often mistreating and neglecting his son.

Errol Graham Musk was born in South Africa in 1945. By many, he is considered somewhat of a mystery. Not only do the other Musk family members rarely talk about him publicly, but he is also a very private man. Other details about Errol's life and work are very much shrouded in rumor. What we do know for sure is that Errol made a lot of money through his work as an engineer, and his relationship with his family became increasingly contentious.

Errol has been described as a manipulating, abusive, and (in general) a terrible man by his children. A gun owner who wasn't afraid to defend his property, Errol shot and killed three people after they tried to invade his house in Johannesburg. However, he was later acquitted of the crime since it was considered self-defense. Little is publicly known about his private life, due to him keeping a low profile. Although, what we do know is that after his divorce from Elon's mother, he married a widowed mother of three children, Heide Bezuidenhout however they too would follow the same path of separation. Only several years later, news broke that Errol was expecting a baby together with his stepdaughter, Jana Bezuidenhout. The scandal took the world by storm and was soon followed by rumors of a second child on its way not long after.

His mother, Maye Musk, was born in Canada on April 19, 1948, and worked as a model for over 50 years. She has a twin sister along with four other siblings. After moving to South Africa with her family in 1950, Maye would frequently enjoy adventurous trips around the country, enabled by her parents' successful chiropractic practice. Maye started modeling when she was only 15 years old, and married young to Errol Musk. According to Maye, the marriage proved to be a problem from the beginning. From the wedding day onward, Errol became physically and emotionally aggressive towards her. Even though she was abused, the law at the time did not favor the women in divorce, so Maye remained married until a change in legislation allowed for a more favorable separation.

After legislative changes in 1979 and the completion of her divorce in 1980, Maye found herself alone, trying to support and raise her children. In an attempt to pay the bills and overcome the financial difficulty her family faced, she started working more while studying to earn a degree as a nutritionist, for which she today holds two Master's degrees. Her work as a dietician helped provide for her family when modeling gigs were scarce. To this day, Maye is still a highly requested supermodel, even at the age of 73—she is the oldest model signed for the makeup brand *Cover Girl*.

Errol and Maye's three children were born in a period of three years. Elon's younger brother, Kimbal, was born on September 20, 1972 in Pretoria. Although not as well-known as his older brother, Kimbal is similarly successful in his own endeavors. After Elon left to live with his father in 1981, Kimbal followed him four years later, in 1985. Kimbal is currently an entrepreneur in the restaurant business and a philanthropist focused on helping improve the world through alternative initiatives, such as the Green DAO. He is constantly seen wearing his cowboy hat in pictures, which has become his trademark. He was Elon's partner when they founded Zip2, but has since moved away from the field of technology. However, he still currently serves as a board member of Tesla, SpaceX, and Chipotle—granting him his millionaire status.

As a trained chef, Kimbal Musk has developed several initiatives to help children eat better and provide healthy food to others, particularly in schools. As one of the co-founders of the Kitchen Restaurant Group collective, Kimbal created the Big Green non-profit organization in 2011. The initiative strives to teach children where their food comes from and how they can grow it by implementing gardens in American schools. He also started an indoor farming initiative, the Square Roots Urban Growers, in 2016.

Elon's younger sister, Tosca Musk, was born on July 20, 1974. An incredibly talented and successful woman in the filmmaking business, Tosca has years of experience and education under her belt. After graduating from the University of British Columbia, she went on to work as an executive producer, producer, writer, and director of films, television shows, and other internet content. She began her career in 1999, and has participated in many productions since then. In 2017, Tosca co-founded a streaming service, Passionflix, focused on romantic films and stories with fairytale endings.

Her first movie with Musk Entertainment, a company she founded and currently owns, aired in 2001, for which Elon acted as an executive producer. Tosca was named one of the 100 names to be aware of on internet television in 2006 for her work in *Tiki Bar TV*, a program that ran from 2005 to 2009. The following year, in 2007, her movie *Simple Things* won several awards. She is currently not married and has two children.

Elon is constantly seen in the media with his brother, sister, and mother—taking them to parties and premieres. According to interviews in which they have spoken about the subject, Kimbal, Tosca, and his mother have stated that they just want to be there for him and to show their support. They generally do not talk about business, respecting the importance of family time when they are in each other's company. The family have reportedly claimed that they do not interfere with the billionaire's business decisions and what he decides to do with his money. In fact, money is something that the Musk family is not very keen on talking about, particularly if it relates to Elon.

Early Education

Aside from his tense relationship with his father, Elon faced many challenges as a young boy. Describing himself as a bookworm who loved to read anything he could get his hands on, Elon has noted that his innate nerdiness did not make his life easy. In several interviews, Elon has claimed that he was bullied in school. His childhood was not a happy time for him, during which he suffered isolation and loneliness. It's no wonder that Elon remembers his childhood without nostalgic tinted glasses.

A large part of this may be due to the fact young Elon probably struggled with Asperger's Syndrome. In recent interviews, Elon Musk admitted that a large part of his bullying stemmed from his habit of correcting people who made incorrect statements. This is a characteristic that he still has to this day. As a child, his very matter-of-fact communication and sharp speech bothered his schoolmates, leading to ongoing bullying.

Waterkloof House Preparatory School, the private school Elon attended, did not provide a great environment for the young boy, who was known to be a geek and an introvert who adored science fiction. He claims that his worst years to date were those in school—times which he frequently describes as a *nightmare* and an *unfortunate time*. During this period, there was an incident in which one of his closest friends tricked him so that older kids could beat him. Elon was also pushed down a flight of stairs by school colleagues, which led to his hospitalization.

However, as a young boy, Elon found happiness exploring and discovering technology and physics. Whenever he visited someone else's home, he'd usually end up finding their bookshelf and diving deep into a book.

In an interview with *Forbes*, decades later, Errol recalled the moment he realized that Elon's future as a genius and innovator was decided. Hearing about a certain lecture at a nearby university, Elon begged for permission to go and listen about computing. The young boy was allowed to go as long as he dressed well and promised to sit quietly. Errol dropped his son off in the lecture hall. Leaving with Kimbal to get a hamburger, Errol returned after three hours to discover no sign of Elon. After waiting for what felt like forever, Errol entered the lecture hall to find Elon comfortably involved in a group of professors and students, conversing about computers. "When I walked up," Errol said, "one of these professors, who didn't even bother to introduce himself, said this boy needs to get his hands on one of these computers". From then on, Elon's interests were firmly fixed within the technological fields.

Soon after, when he turned 15, Elon went through a growth spurt and shed his awkward, vulnerable self. Enrolling in karate and wrestling classes, Elon soon learned how to defend himself from the bullies, who finally left him alone. Graduating from Pretoria Boys High School in 1989, Elon enrolled in the University of Pretoria, which he attended for five months before he was confronted with two choices: to remain with his father or emigrate with his mother to Canada. It was an easy choice. Living with Errol Musk had become impossible, due to his father's life choices and behavior. The opportunity to study at a prestigious university beckoned. Furthermore, the cultural and political scene in South Africa was becoming more difficult to navigate for Elon.

South Africa, a country wracked with racial prejudice and societal instability, still relied on the apartheid system, which Elon was aware of and very much disliked. According to an expose by the *New York Times*, unlike many other liberal families, Elon had built genuine relationships with black friends his age. When a stranger used a racial slur toward his friend Mashudu, Elon spoke up and was then bullied for it. Later, when Mashudu sadly died in a car accident, Elon was one of the few white people to show up at the funeral, something relatively unheard of at the time. In order to avoid mandatory service in the army, remain with his family, and study in North America, Elon elected to emigrate to Canada with his mother.

Starting Life Anew

It was in Canada in 1990 that Elon Musk was accepted to Queen's University in Kingston, Ontario. He studied at the Canadian university for two years before transferring to the University of Pennsylvania in the United States. Some of the closest friends Elon has today are from his time at Queens University. Justine Wilson, his first wife and mother of six of his children, met Elon while attending Queen's university.

When he transferred in 1992 to the United States, Elon decided that he would pursue a bachelor's degree in economics and a second degree in physics. One of the main reasons, according to him, for choosing physics as a major is because the subject constantly seeks out facts in pursuit of the truth—something he is obsessed with. He successfully graduated university in 1995 with both degrees. During his studies, in 1994, he held two internships—in the Pinnacle Research Institute and in a startup called Rocket Science Games. Both companies were located in California, which would later become the entrepreneur's home.

After completing his education at the University of Pennsylvania, Elon moved to California, where he was accepted at Stanford University to study for a Doctorate in physics. It did not last long though. Almost overnight, the internet boom exploded, catching Elon's interest. The young soon-to-be entrepreneur scrambled to launch his first company with his brother, Kimbal.

Even after attending two universities and obtaining two degrees, Elon claims today that studies are not essential to hold a job at one of his companies. Although he targets students and successful businessmen and women to join him in his ventures, having a degree is not mandatory for work in his companies. Contrary to the usual standards of many corporations, Elon values experience and personality. He believes that the university is not a place where you go to learn but to do homework, to build networks, and to have fun. Still, Elon describes his university years as being the most enjoyable time of his life.

Chapter 2:

From Nerd to Billionaire

Persistence is very important. You should not give up unless you are forced to give up.

–Elon Musk

Without financial backing from family and burdened with massive student debt after his years spent in university, Elon launched himself into a Stanford doctorate program. However, the chance to capitalize on the internet boom of the '90s was too good to pass up. With his programming knowledge, the young and enthusiastic man decided to create his own company and try to benefit from the newly commercialized computing utility—and that was just the beginning of his successful career. Since then, Elon has continuously developed and procured other companies, consequently building his career and leading him to become the richest man in the world.

Elon Musk's fortune today directly results from his investments, hard work, and passion for his projects. At most, Elon may have received from his estranged father a supposed $28,000 dollar loan. In his interview with *The Rolling Stones*, Elon asserted his independence, stating:

> He [Errol Musk] was irrelevant. He paid nothing for college. My brother and I paid for college through scholarships, loans and working two jobs simultaneously. The funding we raised for our first company came from a small group of random angel investors in Silicon Valley.

Whether this is true or not, it is clear that a lion's share of Elon's success was due to his and Kimbal's hard work. As seen by his long history of workaholic tendencies and hyper focused attention on projects, Elon's life exemplifies the entrepreneurial visionary, albeit an extravagant and demanding one.

Zip2

Zip2 was founded in 1995 by Elon, his brother Kimbal, and an associate, Greg Kouri. They each had very little money to invest, and sacrificed a lot to front the funding. Elon recalls the long hours they spent programming and starting up the business. During that time, sharing a computer with his brother since industry grade computers, were quite expensive. The young entrepreneur lived in the company's office, slept on the couch, and showered at the YMCA. In a commencement speech at the University of Southern California, Elon also said, "I briefly had a girlfriend in that period and in order to be with me she'd have to sleep in the office". Such cramped quarters, restrictive lifestyle, and long work hours might have driven anyone crazy, but it defined Elon's work ethic moving forward. Could they have found more comfortable living quarters? Perhaps. However, Elon wanted to direct all of the available money to the project, and it seemed to him that renting a house at the time was unnecessarily wasteful when they could focus on redirecting those funds to advance the business.

Since there was so little money available, they used the funds they acquired from several small angel investors to raise the approximate $200,000 they needed, with a little more than 10% coming from Errol Musk, his father. Initially known as The Global Link Information Network, the company wanted to enable businesses to have an online presence to take advantage of the upcoming internet sensation. The Palo Alto-based company kept growing during the day; and, every night, while it was offline, Elon continued to optimize its programming and code. Struggling to close contracts and continue working, in 1996, the company received approximately $3 million in investments and changed its name to Zip2. With the new funding, Elon and his brother were forced to forfeit the majority ownership of their startup, although they still remained on the company's board of directors.

With new investment, the Zip2's new owners changed its main objective, and it soon started developing online city guides for newspapers such as *The New York Times* and *Chicago Tribune*. Elon remained as the company's CTO and held 7% of the shares. With the number of subscribers growing, and the business starting to close contracts with prominent newspapers, it began to capture the attention of larger companies. It finally received an offer and was sold for $307 million in 1999 to Compaq, for which Elon received his share of $22 million. However, rather than enjoying a vacation and his new millionaire status, Elon started a new venture that would drastically impact the financial and online landscape, even today.

X.com and PayPal

As soon as he received the finances for his share in Zip2, Elon decided he would invest in internet banking, as he believed that would be the future of the financial system. X.com provided people with the facility to access online financial services, such as payments and money transfers. Not only was Elon right, but he succeeded in gaining clientele for his new endeavor. Thousands of clients joined the platform. As co-founder, Elon oversaw operations during the first couple of months. The young programmer wasn't finished yet, though.

The popularity and resulting success of X.com was such that, one year later, the company acquired a software company called Confinity Inc., leading to the creation of today's most popular payment giant, PayPal. With the ability to transfer money and make payments by using an e-mail address, the company offered immense financial flexibility and potential to companies competing on the internet. Elon believed in the person-to-person transfer service they could provide and abandoned the idea of an institutionalized online bank.

eBay, in particular, benefited heavily from, and fed into, the success of PayPal. The two sites grew in popularity together. Due to the convenience and breadth of eBay's offerings, where millions of purchases are made every day, PayPal became a household name for individual and commercial financial transactions. The success led to PayPal becoming an open capital company in 2002. However, things would take a sharp turn for the budding entrepreneur.

While he was en-route to Australia for his honeymoon with Justine, he was removed from the position of the company's CEO by the board and replaced by Peter Thiel. Upon landing, Elon received the news and was forced to return to America immediately, canceling his honeymoon.

Regardless of these sudden demands on his time and unplanned changes, Elon continued to ride the wave of success. Although he had been demoted from the role of CEO, he still owned approximately 11% of the company, and when it was sold to eBay in 2002, he received $180 million from the $1.5 billion deal. Elon recalls those early days, saying, "I did reasonably well from PayPal. I was the largest shareholder in the company and we were acquired for about a billion and a half in stock and then the stock doubled".

At this point, it seemed as though Elon had gained enough wealth to pursue whatever it was he truly desired. As a result, he simply moved on to his next business venture and has never returned to working with internet solutions... until 2022. Accounts say that in 2017, the entrepreneur repurchased the domain X.com from PayPal for an undisclosed amount, claiming sentimental reasons. However, when attempting to access the domain by typing it in a web browser, it returns a blank page with the letter x on the top-left corner. What Elon intends to do with the relic of his starter domain remains to be seen.

The Price of a Vacation

After starting and developing his second company successfully, Elon might be forgiven for putting his feet up and enjoying some of his well-earned money. Intending to enjoy his now delayed honeymoon with Justine, Elon and his wife traveled to South Africa in December 2000. Upon their return home, however, the happy romantic mood was spoiled. Elon had become severely ill, and his condition worsened through the month of January. Elon contracted a particularly dangerous type of malaria that nearly ended his life. According to a tweet Elon shared, he said:

> I was first misdiagnosed at Stanford Hospital with viral meningitis, then again misdiagnosed at Sequoia. A visiting doc from San Jose General saw my charts & sent me to the ICU immediately. I was ~36 hours from being unrecoverable. So I take expert advice with a grain of salt...

This skepticism would crop up again later on in life, and it would determine his cautious view on taking breaks. Looking back at his experience, Elon once joked, "That's my lesson for taking a vacation. Vacations will kill you". While it might have been treated as a joke, Elon's work ethic denied him long vacations for over a decade.

Following his struggle with malaria, Elon threw himself back into his work, without looking back. As someone who always seeks new possibilities and the truth behind the world, the entrepreneur wasn't content with just sitting around and spending his money. It wasn't that Elon didn't enjoy relaxing every now and then with a book, but he needed a new challenge.

In an interview with the Computer History Museum, Elon noted that the outcome of his work with PayPal left him very well off. He added:

> So yeah, I did reasonably well, but the idea of lying on a beach as my main thing, just sounds like the worst — it sounds horrible to me. I would go bonkers. I would have to be on serious drugs. I'd be super-duper bored. I like high intensity.

The question was... what next would catch Elon Musk's interest?

Chapter 3:

Expanding Horizons

Going from PayPal, I thought well, what are some of the other problems that are likely to most affect the future of humanity? Not from the perspective, "what's the best way to make money," which is okay, but, it was really "what do I think is going to most affect the future of humanity."

–Elon Musk

Space exploration has captured the imagination of many innovators, inventors, scientists, and entrepreneurs. However, as NASA's historical struggle with finances proves, the field of space exploration often requires long-term investment, with little profit shown for the astronomical start-up costs. Inspired by science fiction, a genre Elon has always enjoyed reading, the entrepreneur recognized that humanity needed to blast towards the next step in space exploration.

Believing that human civilization is cyclical, Elon hoped that he would be able to take advantage of this window of opportunity. In an interview with *The Guardian*, Elon stated that this was "the first time in 4.5 billion years where it's been possible for humanity to extend life beyond Earth" and that it would "be wise to act while the window was open and not count on the fact it will be open a long time". With this self-appointed task, Elon began to work on two projects that he believed would bring about a more positive future for humanity: SpaceX and Tesla.

SpaceX and Starlink

Crazy, naive, bold—these are just some of the words that were used to describe Elon Musk when he announced that he was going to focus on the colonization of Mars through his collaboration with the Mars Society. He believed that the development and human settlement of Mars should be one of the next strides in space exploration. Determined to test his thesis, Elon began to plan how to jumpstart this new project. But first, he needed to find a way to make space travel more affordable than it was.

Initially, Elon considered launching a glass-enclosed greenhouse to the red planet. A project titled Mars Oasis. Seeds would be embedded in dehydrated nutrient pods and released upon landing, sending back vital imaging and data points that would reveal a lot about the viability of transporting life to Mars. However, in the United States, the cost to acquire a launch vehicle far exceeded his budget; around $65 million per rocket. In a turn of events, Elon would find himself in Russia attempting to purchase used rockets that he could repurpose and study. Unfortunately, it only resulted in more rejection. In one meeting, Elon even got spat on by one of the participants. None of that phased the would-be rocketeer. He was determined to make this project succeed.

Eventually, Musk would attempt to negotiate the price of a pair of de-nuked Russian ICBM's but the deal never eventuated. With cost and limitations of technology remaining to be a real impediment, Elon decided that he would create his own company dedicated to the manufacturing of rockets. Aiming to make space travel commercial, and win contracts with the government at the same time, Elon focused on self-sustainable methods. The entrepreneur was finally able to develop a prototype in the early 2000s. For this, he invested nothing less than $100 million of his own money to establish the Space Exploration Technologies Corp., also known as SpaceX, in 2002.

As both the company's CTO and Lead Engineer, Elon hired numerous professionals from the aerospace industry to work on this innovative project. The idea of affordable rockets meant, to most, that they were unreliable. However, Elon once again decided to prove to people that they were wrong. His first launch was with the Falcon 1, in 2006. Falcon 1 was a failure, with the rocket failing to launch and subsequently catching alight. Elon refused to quit. He kept on persisting until 2007 and 2008, although some issues were still at large, and the company was unable to reach space orbit successfully. Remarkably, in September 2008, SpaceX performed its first successful mission, sending a fuel rocket to outer space.

Soon after, the company was awarded contracts with NASA and the U.S. government for $1.6 billion. By 2009, SpaceX had already launched a series of space Falcon rockets, named after the infamous *Star Wars* ship, the Millennium Falcon. Although the first launches failed, the following rocket launches were more successful. With each success and failure, Elon and SpaceX continued to learn and re-iterate with each rocket. In 2010, he launched the Falcon 9; a rocket with 9 engines. Unlike traditional space ships, the Falcon 9 was equipped with the ability to land vertically after releasing its second stage and payload at a certain altitude. By January 24th 2021, the pioneer rocket had set the record for the most satellites launched by a single rocket, carrying upwards of 143 into orbit.

Another SpaceX rocket, the Dragon, proved also to be a great success, leading to the eventual dispatch of food to the International Space

Station (ISS) in 2012. The Dragon rocket has already had 34 launches and visited the ISS 30 times. It can carry up to seven passengers to Earth's orbit and has become the first private spacecraft to take humans to the space station. Today, SpaceX is most famous for being the first private company to reach Earth's orbit and, subsequently, to connect with the ISS.

Not content with redesigning space rocketry alone, SpaceX went to work on developing other important space solutions, such as innovating on the traditional astronaut spacesuits. Instead of remaining content with the bulky suits that NASA developed decades ago, Elon and SpaceX have reinvented the spacesuit, bringing it to the 21st century. The sleek new suits debuted in 2020. Worn by veteran astronauts Doug Hurley and Bob Behnken, the suits received a five star rating thanks to their minimalist design, high technology, and convenience. Even the gloves received special attention, designed to work with the spaceship's touch screens. The suits, like much of SpaceX's work, speaks to Elon's vision for innovation and commitment to excellence.

Another one of SpaceX's most notable projects is Project Starlink, which is a satellite constellation internet system. These satellites, carried to space, soar past in a low orbit around Earth. Providing internet by satellite to the countries where it is licensed, the SpaceX product is available in 32 countries. While the Starlink routers and subscriptions are expensive, with the monthly fees sometimes rising over $100, Starlink's services are second to none due to the technological advancements SpaceX achieved with their hardware.

According to the company's founder, it is expected that the project's reach will be global by August 2022. Currently 98% of the users are from the Western hemisphere, but Elon aims to extend services internationally. Still, Starlink has come a long way since its inception. The project started in 2015, and the first satellites moved into orbit in 2018. The company started to provide internet services in 2021 after it had already launched over 2,000 satellites into orbit. Although the idea was to serve any client desiring a fast connection to the internet, it was specifically designed for those who live in rural areas and do not have easy access to the service through conventional channels.

Despite SpaceX's intentions to provide stable internet services for anyone anywhere, it has received its share of criticism from the scientific community. According to astronomers, the light emitted by these satellites in the Earth's orbit can make it harder to make scientific observations of the stars, planets, and space in general. Criticism, though, is Elon's best friend. Thanks to input from his fellow scientists and academics, Elon and SpaceX aim to continue innovating Spacelink technology. According to Elon, SpaceX has already begun testing new products which emit less light and are already undertaking the necessary studies to transform Starlink into a more efficient and environmentally friendly product. They have also claimed that they are developing a system to enable the satellites to float out of orbit and into space once their life cycle is over. As a result, SpaceX products will not contribute to the ongoing problem of cyber-trash currently cluttering Earth's orbit.

Tesla

Although many people might think that Elon Musk was one of Tesla's original co-founders, the first version of the company, Tesla Motors, was in fact founded in 2003 by Martin Eberhard and Marc Tarpenning. After meeting Elon and connecting with the engineer, the three men collaborated and created the company we know as Tesla today. Elon's involvement with the company began in 2004, when he invested $30 million in the company's first investment round and helped develop what would become to be known as the Tesla car. As the leading investor and largest shareholder, Elon skyrocketed to the position of chairman. According to Elon, when he first invested in the company, there were no employees and there was no product. Since Elon was involved with Tesla from the ground up, the entrepreneurial engineer considered himself a co-founder of the company.

Disagreements between Elon Musk and Eberhard, due to this information, eventually led to the former leaving the company and filing a lawsuit later against Tesla. In the lawsuit, Eberhard claimed that he was forced out of his own company, and that he and his former partner were the only ones who had the right to hold the title of company founders. The case was ultimately settled, and Elon's claim of co-founder was upheld in court. It was estimated that, by that time, Elon had sunk over $70 million of his own money into the car company.

When asked during a TED Talk in 2022 if there was anything he could change in the past few years, Elon claimed that he would change the way he invested in Tesla. He said that he would instead have started a new company with one of the other co-founders rather than investing in the existing company, which would have made life easier for him and the whole process less stressful.

Since then, Elon has remained Tesla's CEO and leading product architect. He held these positions since 2008, when the company's first car, the Roadster, started manufacturing. The Roadster was considered an innovative car because it was the first of the electric models to travel almost 400 kilometers (248.5 miles) with a single charge. It also achieved a speed of up to 200 kilometers (124 miles) an hour—something that had never happened before in the industry. However, the price was considered a luxury for many. The Roadster was sold for a little over $100,000, taxes included. Since then, Tesla has improved its models, particularly focusing on ways to reduce the time required to recharge the vehicles. They have also significantly increased charging locations for the vehicles which are called Superchargers. As Tesla proved that its cars could very much compete with regular gas-fueled cars, sales began to increase.

Under Elon's directive, the company altered its name from Tesla Motors to Tesla, Inc. as he decided to increase the company's scope. Elon hoped to develop new energy projects based on solar energy. In order to achieve this goal, one of Elon's strategies involved the acquisition of the second largest provider of solar system energy in the United States: SolarCity. During this time, Elon saw Tesla as a force for positive environmental impact, a goal he continues to pursue today.

The acquisition of the New York company led to the construction of Tesla's second factory in the United States. This time, Elon built a factory in Buffalo, New York, which has become known as the Gigafactory 2. The first Gigafactory had been built in Nevada. As Tesla innovated on production, the Tesla car model eventually declined in price, leading to another surge in sales. Elon's dream of making an environmentally sustainable car more affordable to the general public was finally coming true.

In 2019, Elon announced and fulfilled his promise to release Tesla's patents as part of an initiative to combat climate change and global warming. The entrepreneur has stated multiple times that he believes that the efforts towards dealing with ongoing environmental problems are not dealing with the issues quickly enough. He has been very vocal about his dislike of the increasing number of fuel-enabled vehicles available in the market, polluting the air. With this decision, he hopes that the big car manufacturers will use Tesla's patents and develop more clean fuel transportation in order to protect the planet.

Neuralink

What if we could download skills and information instantly into our brain? Looking to the future, Elon and a band of partners were interested in answering such a question. It prompted one of Elon's most controversial initiatives: Neuralink. Co-founded by the billionaire alongside eight other scientists and engineers in 2016, the Neuralink team began to explore the idea of integrating artificial intelligence with the human brain. According to the company's proposal, this was to be done by implanting devices in the human brain, leading to a merger between humans and machines and advancing artificial intelligence.

Elon and his co-founders shared their hope that the program would one day help disabled people or even cure patients of brain-related diseases, such as Parkinson's. Although this might seem like science fiction to many, the South African mogul believes that it is within the realm of possibility and that it will ultimately benefit a large portion of the population. Elon's claims have widely resonated in the medical community, which has claimed there is still a significant trail to walk through to understand the brain, and that these affirmations cannot be made without prior study. Thankfully, Elon and his co-founders represented a variety of scientific fields including neuroscience, biochemistry and robotics. It would only be a matter of time before Neuralink would find a specific goal to focus on. Time will tell.

The Boring Company

Tired of the Los Angeles Traffic system, Elon Musk announced the creation of The Boring Company in 2016. Initially, the company was simply a division within SpaceX, focusing on the task of boring tunnels in the city to improve traffic. The small company's first project began in 2017 that involved drilling a hole within SpaceX's premises.

Having gotten permission to drill, Elon dove into action. Within a matter of hours, the employee parking lot was emptied and, by the weekend, working round the clock, a large hole had already been bored out. In 2018, the division was promoted to its own separate company, with Elon retaining 90% of the shares.

The company has already completed two tunnels—one in Hawthorne, Los Angeles, and another in Las Vegas. The Los Angeles tunnel, dug in 2018, was 1.14 miles long. The Las Vegas tunnel, located beneath the Las Vegas Convention Center (LVCC), ended up around 1.7 miles long. During these early years, whether these tunnels would prove a success was widely debated by mainstream media. Still, Elon believed that The Boring Company's tunnels would provide low-cost transportation and relieve some of the traffic congestion in America's main cities. The tunnels were specifically designed to be used by Tesla cars, which will operate at high speed, delivering those to the requested destination safely and speedily. Elon also claimed that, by using tunnels, there would be less use of the surface, thus making cities more beautiful and lowering sound pollution, such as vibrations and noise.

While Elon tackled the problem of traffic, armchair critics emerged to raise questions about The Boring Company's goals and techniques. Other engineering experts voiced critiques regarding the tunnels. They are unsure how the company could dig and ventilate the tunnels at such a low cost, compared to the tunnels that already exist today. Elon refused to back down. Elon pointed out that there would be no need for a massive ventilation system. Since the tunnels were primarily designed to be used only by electric cars, there were no concerns over eliminating carbon monoxide. As a result, the lack of additional systems for the tunnel lowered production costs.

Just the Beginning

After PayPal, jumping into space exploration, encouraging the electric car industry, creating brain chips, and digging long tunnels under bustling cities, you would be right in thinking Elon would have had his serve of challenges and setbacks. Still, during these years, Elon never stopped working, refusing to give up. Instead, he embraced the challenges and critiques, learning from each mistake. His tenacity and passion for truth and practical solutions had already begun to show results, but Elon was only just getting started.

Chapter 4:

The Man Behind the Fame

I would just question things... It would infuriate my parents... That I wouldn't just believe them when they said something 'cause I'd ask them why. And then I'd consider whether that response made sense given everything else I knew.

–Elon Musk

What makes a genius? What kind of person can reach the hall of fame? Which qualities are needed in order to make it to the top? Historically, men of power required charisma, family money, or connections. For Elon Musk, a self-made billionaire, his personality traits and qualities often undercut his skills as much as they birthed them. While his early years of social isolation and introversion made it difficult for him to build relationships, he slowly transformed into an intelligent, risk-taking young man. Later on, Elon began to accept his eccentricities and turn them into strengths, allowing for his straight-talking, dead-pan presentation to set his leadership style and management approaches apart from other popular companies. In turn, his complex motivations and unique personality have driven Elon to tackle some of the world's largest problems in his own special manner.

The Personality of a Genius

From the beginning, Elon Musk has described himself as a bookish nerd who preferred to spend time alone pondering the truths of the universe, reading books on topics that caught his interest, or solving a complex problem. As a child, his love of reading stemmed largely from social discomfort, thanks to his struggle to read social cues. When talking about his journey to maturity, Elon admitted, "I would just tend to take things very literally, just the words as spoken were exactly what they meant". Over time, thanks to his thirst for knowledge, the huge amount of time he spent reading, and an assortment of films, Elon began to understand how to communicate with people better. He realized how to guess at the hidden meaning behind people's speech.

For some, these characteristics might sound familiar. The inability to read social cues or struggling to process the external world are common signs of Asperger's Syndrome, which is now considered to be a form of the Autistic Spectrum Disorder (ASD). During a TED talk, Elon was asked whether his Asperger's encouraged him to process the world inwardly on a deeper level. Elon agreed, but he also pointed out that his experience of Asperger's gave him the ability to focus deeply on programming and technological development. Explaining that he found programming all night rewarding, Elon admitted that perhaps his interests were not normal. "I think most people don't enjoy typing strange symbols into a computer by themselves all night," he said. "They think that's not fun, but I thought it was. I really liked it".

Elon's initial social awkwardness, combined with his love for detail-oriented, independent work, drove him to isolated studies. As a result, while many people see Elon Musk as an entrepreneur, a meme lord, or a practical engineer, there is a hidden side to Elon that is very introspective. When asked what kinds of truth he was interested in seeking out, Elon shared that cosmic, philosophical truths, as much as scientific data, took up a lot of his time and focus. When he was young, he even dove into more philosophical exploration, trying to probe the meaning of life. Reading books on philosophy, as well as religious texts, Elon found himself growing increasingly depressed. "I got into the German philosophers," he said, "which is definitely not wise if you're a young teenager, I have to say, a bit dark, a much better read as an adult". From that time onward, Elon focused on more practical pursuits, which would drive his entrepreneurial projects.

Since Elon has the ability to focus deeply for long periods of time, sometimes at the cost of his physical and mental health, his work has flourished even during very stressful times. Thanks to his perseverance, hard work, and innovative eye, Elon has been able to make the seemingly impossible—possible. However, Musk's ability to drive himself doesn't always translate well when it comes to managing others.

Managerial Characteristics

While he is seen as a visionary and an innovative leader, Elon Musk's managerial style has been constantly criticized by former employees and business partners. It may be difficult to figure out which criticism is fair or not; but, given how far Elon is willing to drive himself in pursuit of truth, you can see how he might be a tough, but fair, manager. In previous interviews, the billionaire has described himself not as a micromanager but as a *nano-manager*, constantly placing his employees under pressure. He is said to have pushed his employees to go beyond their limits on multiple occasions, challenging them to do greater and better things.

According to some accounts, Elon maintains high standards, firing employees who underperform and confronting his staff when projects go off the rails. He is known to seek open and honest communication, bypassing bureaucratic systems. This has resulted in middle management feeling belittled, or made redundant. Still, by telling employees that complaints and issues should be directed to him first, Elon offers his employees an open door for dealing with problems or obstacles instantly. As someone who appreciates working with both positive and negative feedback, Elon prioritizes the easy flow of information within his teams. This also means that he considers meetings a waste of time and counter-productive, unless dealing with an urgent matter.

Since Elon has his own way of dealing with conflict and issues that arise, new employees might struggle with acclimating to his leadership style and work environment. Former employees claim that, although Elon was an inspiring person to work with, most of them would never work in a company run by him again due to the considerable amount of stress and demands that come along with the job. Elon is painted as an inspiring, but scary, leader who can take those who work with him to the limit—of both their abilities and mental sanity.

When working with Elon, you must be prepared to approach all work with a scientific scope. Just as you would develop a scientific thesis, you have to be able to identify the problem, build a hypothesis, do the research, present the findings, and then work on the results to achieve the goal. While this positively impacts his companies' general conclusions and developments, leading to low cost and efficient solutions to problems, it can also stress less creative workers out. The good news, however, is that Elon is not afraid of failure. In fact, the innovative entrepreneur embraces failure as a mode of learning.

Also, Elon Musk is known to have an excellent eye for hiring people. He is said to generally make the correct choice, since hiring the wrong staff can cost a company a lot of money. He also strives to provide a creative, motivational work space for his employees, so that they are as happy as him to come to work. Elon has shown himself to be flexible and open to suggestions to improve the workplace and satisfy employees.

Elon's flexibility is also notable when he is working on a project—he can change his mind mid-project about a particular issue, and change the whole course of the endeavor. He likes people to be proactive and expects problems to be foreseen before they arise, another positive impact of his leadership in his companies. Most of the time, this technique is efficient and leads to promising results, especially when considering the short deadlines the entrepreneur sets for his teams.

Unlike other CEOs, Elon leads by example, not content with putting his feet up and smoking cigars in his office suite. Instead, he is described as a dedicated businessman, who works long hours when he is not with his family. Going so far as to sleep on the office couch for nights on end, Elon will remain at his office or factory 24/7 until a problem is solved. When sharing stories about his time setting up his first Gigafactory, Elon shared his work flow. "I lived in the Fremont and Nevada factories for three years," he said, "fixing that production line, running around like a maniac through every part of that factory, living with the team". Elon did this so that the team, who he knew was going through a difficult time, would see that his commitment was translating into real work. Elon would not require of his team what he himself wasn't ready to commit. In this way, Elon made sure that he wasn't disappearing into an ivory tower, but experiencing the same mental and physical stresses they were enduring. The dedication Elon modeled has inspired many other entrepreneurs and future CEOs. But, it's not just about working hard to make money. The man behind the billion-dollar companies has an even more important self-ordained mission which drives him.

The Future of Humanity

In spite of being a man with money, family, and children, Elon Musk considers himself a man on a mission with a purpose. This purpose is to create a better future for humanity, saving it from the results of its worst excesses. Elon wants to solve traffic, cure diseases, and bring new, more efficient solutions to current difficulties to provide a better life for all of human civilization. As a father, Elon hopes to set an example for his children and future generations of young people. Since he isn't going to live forever, Elon not only hopes to create a world that will nurture and sustain civilization going forward, but he also wishes to preserve Earth and its natural environment for the physical and mental health of humanity.

As the billionaire has claimed in several interviews, he is profoundly worried about the future of humanity, for a variety of reasons. For starters, Elon is concerned about the decreasing number of children being born in more recent years. According to him, the rate of children being born is unsustainable to maintain civilization in the future, which he is deeply concerned about. Secondly, Elon has noted on several occasions that the development of AI might cause problems its creators have never intended. He urged governments to place restrictions on AI and facilitated the launch of an open-source, ethical AI under OpenAI. However, Elon's real concern is climate change. With the creation of Tesla, he hoped to encourage sustainability, when it comes to the use of natural resources. For example, the entrepreneur wishes to build more efficient technology for clean energy. Echoing the sentiment of many in the scientific community, Elon has stated multiple times that alternative energy sources must be pursued in more meaningful, practical ways.

It is true that humanity has already taken a huge step in supporting the natural world of planet Earth by using electric cars and developing alternative energy sources, such as solar or wind power technology. However, for Elon, that is only the beginning. Since the United States still largely relies on oil and coal for energy, Elon hopes to challenge himself and other entrepreneurs and corporations to compete to produce a cheap and efficient means of creating energy. Elon even extended his challenge to other companies and innovators, which will grant the winner $100 million if they can develop an efficient, and large-scale, solution to remove carbon monoxide from the atmosphere.

According to Elon, the first steps to shaping a better planet have been taken, although much more needs to be done. His vision is that, in 2050, the world will be in a better state—where cars with fossil fuels will be close to nonexistent. By that time, Elon hopes that the majority of energy consumption will come from renewable and sustainable sources, from solar to nuclear energy. Elon claims that he will do everything within his control to make it happen—starting with the fabrication of more efficient batteries in his own factories.

Listening to Elon's ideas, it is clear to see that he is more than just a businessman or tech nerd. His personality and dreams drive him to the limit, keeping him busy with 80-hour work weeks, as he models for his employees his own commitment. Instead of just talking about helping our planet or creating a better future for the generations to come, Elon is putting his money, and his mind, on the line. It is no wonder, then, that Elon Musk has become a subject for scrutiny and an inspiration for scientists and business people around the world.

Chapter 5:

Finding Balance

*I will never be happy without having someone.
Going to sleep alone kills me.*

–Elon Musk

In the later months of 2021, Elon Musk shared an update on his SpaceX project, Starship, at a virtual presentation from within his home. Dressed casually in a suit jacket and loose white shirt, Elon's laidback look was a far cry from his usual buttoned-up self. Even more charming was the sight of his young son, X AE A-Xii, sitting on his knee. While Elon attempted to talk about his fully reusable spacecraft, X AE A-Xii stole the spotlight—waving his arms, saying "Hi!" and soaking up the attention from the Zoom callers.

Eventually, Elon's son was whisked out of the room, but the sneak peek into the entrepreneur's business life revealed the dichotomy that is Elon Musk. On the one hand, we know Elon for his brutally honest statements, skeptical queries, and memes. On the other hand, Elon plays other familial roles—son, brother, husband, and father.

Known worldwide for his business endeavors, Elon's personal life usually ends up on the back pages of the tabloids. Although the media is always looking for more details about Elon's personal life, the engineer and businessman has repeatedly stated that he desires to keep his private life *private*, regardless of public interest. Once in a while, however, we are offered a glimpse into what the private life and world of Elon Musk might look like. Whether one of his many children are mentioned in an interview article, his current partner announces the arrival of yet another child, or Elon experiences another break up, each piece of information we hear shows that Elon's life, like our own, has its share of ups and downs.

Relationships and Children

Treasuring his private life and respecting the privacy of his ex-partners and children, Elon has yet to confirm or deny many accounts. What we do know is that Elon has had a variety of female partners over the years, which resulted in a total of 8 children. Married and divorced three times, twice to the same woman, Elon clearly places a lot of importance on relationships in his life. In interviews, he claims that he needs to be with someone and that he does not like to be alone, which is something that leaves him miserable. As a result, Elon makes sure to keep his family close, remaining an active figure in his children's lives, while also pursuing love when he can find time in his busy schedule.

From the beginning, Elon valued romance and family. During his time at Queen's University, Elon met his first wife, Justine Wilson, who entered university a year later than the young engineer. Although they enjoyed each other's company, it took a while for the couple to eventually wed. Thanks to issues with PayPal, Elon and Justine's first attempt at a honeymoon failed. The second honeymoon, in late 2000, nearly ended Elon's life due to the malaria attack.

In 2002, Justine's first baby, Nevada Alexander Musk, unfortunately died when he was just ten weeks old of sudden infant death syndrome. Out of grief, Justine pursued more babies, this time with a fertilization clinic. Thanks to the clinic's success, she became the mother of five more of his children. Elon and Justine had twin sons in 2004 (Xavier and Griffin) and triplets in 2006 (Kai, Saxon, and Damian)—all conceived by IVF.

Unfortunately, the couple struggled during their relationship and decided to divorce in 2008. A man who was as equally married to his work and ambitions as he was to his wife, Elon was said to be the dominant partner, making decisions and dedicating himself to work at every available moment. Justine has gone on record, in a piece written for a magazine after the divorce, to say that Elon viewed her as a starter wife. Elon supposedly expected from her something that she could not possibly deliver. Despite attempts at couples therapy, Justine claimed Elon gave up, refusing to return after one session. Allegations that Elon had been unfaithful to her were implied in the same piece, which Elon refuted. Despite the rocky ending to their relationship, Justine and Elon live close to one another, allowing them shared custody of their children. Although they reportedly do not get along well, Justine claims however, that she does not regret the relationship.

The allegations about cheating arose in the revelation of Elon's new relationship. Soon after Justine and Elon's divorce was announced, it was reported that Elon had started dating a British actress, Talulah Riley. Elon has since written a response to Justine's essay, stating he had only just met Talulah after the divorce was filed. Elon and Talulah had only known of each other up until the marriage had ended.

Elon and Talulah were married for the first time in 2010, and all seemed to go well, with the actress having a good relationship with the billionaire's former wife, Justine. However, citing "irreconcilable differences," the couple filed for divorce in 2012—a divorce which Elon announced publicly on Twitter at the time. The relationship wasn't over. In 2013, the couple got back together again and remarried, only to divorce again in 2016. In his 2017 *Rolling Stones* interview, Talulah made an appearance, relaxing with Elon and Neil Strauss—showing that Elon and Talulah are still on good terms and supporting the claim that they remain good friends. The couple did not have any children together.

After his two marriages, Elon has not remarried, although he has gone on to have romantic encounters with other women. He had a brief relationship with actress Amber Heard in 2017 after she divorced fellow actor Johnny Depp. In the same interview with Rolling Stones, it is clear that Elon's emotions were deeply impacted by the break up with Amber Heard.

Elon's pursuit for love and companionship wasn't over, though. In 2018, it was revealed that he was in a relationship with the Canadian singer and songwriter Grimes, with whom he had two children—one boy, AE A-Xii, born in 2020, and a girl, Exa Dark Sideræl Musk, born in 2021. Both AE A-Xii and Exa Dark Sideræl, born through surrogacy, currently remain closer to their mother. Elon pointed out in an interview that Grimes has a bigger role thanks to his children's young age. Later on, as they get older, Elon will be able to build a relationship with his children as he has done before. "If I have a trip for Tesla to China, for example," he said, "I'll bring the kids with me and we'll go see the Great Wall or take the bullet train from Beijing to Xi'an and see the Terracotta Warriors".

Is Elon currently pursuing a relationship? All signs point to 'no.' The same year that his first daughter was born, Elon announced that he was single, although he still cared very much for Grimes. While the two remain friends, Elon is free to find love yet again, elsewhere.

Philanthropy Work

Just as he welcomes privacy regarding his family life, Elon Musk pursues anonymity when it comes to his philanthropic efforts as well. As the founder and president of the Musk Foundation, Elon has gone on record as being very invested in supporting practical approaches to philanthropy. According to the foundation's web page, the Musk Foundation provides grants mostly related to the companies in which Elon has participated or currently acts. They support space exploration, science and engineering education, the development of artificial intelligence, human space exploration, and renewable energy research.

Outside of his foundation, Elon has also granted billions in donations to charities. In 2021, he donated $5.7 billion in Tesla shares to an unnamed non-profit organization. The destination of the funds was untraced. This, of course, led to a lot of speculation as to where Elon's donation had actually gone. *Forbes* initially believed that the money went to a donor-advised fund, which they describe as essentially a philanthropic bank account. These types of accounts can have money in them for years, as the publication explains, without having ever been sent to any kind of charity. Still, according to the magazine, only 20% of Elon's previous donations to charities were made to an account of this type. That is a little over $56 million of the $280 million donated so far to multiple organizations.

Among these organizations are the St. Jude's Children's Hospital, public and private schools and non-profit groups in Texas and California, The Mercatus Center in Northern Virginia, and the Wikimedia Foundation. Since 2012, he has been the signatory of the Giving Pledge, which is a promise made by the world's wealthiest men and women that they will donate the majority of their wealth to charitable causes. There have been comments that the billionaire should be more transparent about which non-profit organizations he gives his money to, including claims made by other wealthy individuals. Elon, however, refuses to participate in the self-aggrandizement that often accompanies charitable donation announcements. Marcius Extavour, the vice president of the carbon removal non-profit XPRIZE, said that Elon was more concerned about how the money he donated was spent rather than how it made him look.

In 2012, alongside the comic *'The Oatmeal'* and Nikola Tesla fanatics raising a whopping $1.37 million, Elon also pledged a million dollars to help restore Wardenclyffe: Tesla's lab on the North Shore of Long Island. Promising to help fund the creation of a Tesla Museum on the property, Elon has shown interest in preserving the heritage of the genius who inspired the name for Elon's car company. Since then, Elon has begun to recognize the influence for good that he has online. As a result, Elon has become more vocal about his charitable missions, no doubt due to his rising concerns over global environmental and societal trends.

Instead, Elon prefers to engage in social media in order to promote charitable causes that have caught his eye. Whether it be to plant trees, end world hunger, or solve the climate crisis, Elon enjoys challenging himself and his audience to practically deal with ongoing global issues. He even created a climate crisis challenge, for which the winner will receive approximately $100 million for helping solve the climate crisis. On another occasion, the billionaire claimed that he would donate $6 billion to help world hunger if it was proven to him how the money would help solve the problem. Instead of simply throwing money at a problem, Elon is committed to long-term positive impact through practical aid and planning.

Vision on Education

Since Elon has a large family and a commitment to future generations, it is no surprise that he has a few opinions about our current educational system. According to Elon, America's educational system is something that could be tremendously improved. This is something he firmly believes, driving him to build his own school. When he was unhappy with his children's education department, the entrepreneur decided to build one for them and his employees' children inside SpaceX's facilities. Does that sound familiar? It should! Whenever Elon identifies a problem or inadequate situation, it is characteristic for the billionaire to arrive at his own solution, creating something better himself. Thus, the founding of Ad Astra commenced.

The school was established in 2014, opening its doors for children from 6–14 years old. Named Ad Astra, the school aims "for the stars." As a result, it approaches education with a more unique focus. With no separation per grade for the children, education at Ad Astra focuses on the strengths of intergenerational learning, pro-active study, and independent problem-solving rather than memorizing information. Since the students are separated by ability instead of age, the learning process focuses on their skills and aptitudes. As a result, the children learn to work with a variety of ages outside the limitations of their peer group. On top of that, there are no grades to evaluate the students.

Of course, the primary focus of studies at Ad Astra are a reflection of what Elon deems to be particularly important for future generations (e.g., math, engineering, and technology). Despite the school's hyper focus on the sciences, the underlying educational philosophy has inspired other educators.

In 2017, an engineer named Chrisman Frank visited the school and recognized that all children should be as enthusiastic about learning as the students were in Ad Astra. Joining forces with Josh Dahn—a former teacher from the school Elon's children used to attend, who was now teaching at Ad Astra—Chrisman Frank launched an online education system in 2020 called *Synthesis*. Providing a subscription service focused on forward-thinking education, the tool targets children from 6–14 years old, with the aim to encourage critical thinking, independent learning, and problem-solving.

Still, thanks to his concern and interest in education, Elon has donated millions of dollars to several schools, in order to improve the programs offered to their students. As a man who places a premium on truth and critical thinking, Elon wishes to revolutionize learning. He hopes that children around the world, regardless of their cultural or economic background, will learn how to think critically, act with confidence, and impact the world positively. This might come as a surprise to people who have only heard Elon's critique of post-secondary education.

When discussing university, Elon has gone on record to say that the post-secondary system needs a lot of reworking. Claiming that today's educational system does not enable students to learn, Elon said that college has become a place to have fun and complete mundane homework instead of pushing yourself and the boundaries of learning. While he had fun at college, Elon admits that much of the knowledge he values today was gained through experiencing life hands-on.

What Spare Time?

After recognizing all of the work that Elon ends up involved in—SpaceX, Tesla, his side interests in education and philanthropy, as well as his family life—it is no wonder that Elon seems to have no spare time. During some periods of his life, Elon would be working upwards of 80 hours a week, resulting in eventual burnout. Recalling those times, Elon says, "There were times when I didn't leave the factory for three or four days - days when I didn't go outside... This has really come at the expense of seeing my kids. And seeing friends".

Since 2018, Elon has been trying to find a better balance between work, sleep, and his personal life. If he isn't working or thinking up solutions to ongoing challenges, Elon will usually spend time with his children and family. During quieter moments, Elon continues to enjoy reading and gaming. More recently, he shared his *Elden Ring* character build on Twitter, revealing that even the super busy billionaire makes time for one of the most hyped video games of all time. Despite the busyness of his life, Elon (of course) somehow finds a few precious seconds to send out a Tweet...or two, or three.

Chapter 6:

Twitter Wars

A social media platform's policies are good if the most extreme 10% on left and right are equally unhappy.

–Elon Musk

A man with so much money and so many controversial opinions is bound to call attention to himself—and that is true about Elon Musk, as well. Whether it be through his social media accounts, interviews, or announcements, it is rare to see the billionaire shy away from hot topics. Many might see his defense of freedom of speech as an excuse for his confrontational communication. Certainly, thanks to Elon's comments in the past, he has faced lawsuits and investigations. A variety of voices, particularly on Twitter, spoke out against Elon's questions, opinions, and statements—most notably during COVID-19. For example, Elon's refusal to close Tesla offices generated a lot of noise from the public.

What is clear is that Musk, a man who is unafraid of saying or tweeting what he thinks, be it good or bad, isn't afraid of confrontation. As a man who does not hold back when issuing an opinion, Elon may often be proven wrong, but he views it as part of his mental processes as he looks at situations in real-time. Since many of his tweets are off-the-cuff reactions, it is often hard to tell if he is joking or being serious. Even when it seems like Elon is just joking around, sometimes he changes his mind and decides to develop a project, as happened with the flamethrower and Tesla Tequila. More recently, Elon has been offered a chance to acquire Twitter, which has led to further controversy.

The Cave Rescue Tweets

In 2018, 12 members of a children's soccer team were trapped in a cave in Thailand. As tensions grew, Elon volunteered to build a mini-submarine to save the stranded boys. Although the equipment was built in record time, the rescue had already started by then, and there was no need to implement Elon's sub. When all of the soccer team had been rescued, one of the rescuers, Vernon Unsworth, claimed that Elon Musk was simply looking for positive publicity. Suggesting that Elon's offer was a publicity stunt, the rescuer said that Elon's idea wouldn't have been effective, anyways.

Elon's reply stunned and shocked quite a few people, referring to Vernon Unsworth as "pedo guy." Vernon Unsworth was understandably upset about being called a pedophile. Furthermore, the Tesla CEO was accused of paying $50,000 to a private investigator to find out information on the diver. Later, the billionaire explained that this term was slang that is commonly used in South Africa. Regardless, Vernon Unsworth decided to file a lawsuit against Elon Musk for defamation. Elon eventually won in court, but later issued an apology on Twitter for insulting the man.

Thoughts on the COVID-19 Pandemic

While the world struggled to comprehend the impact, origins, and solutions to the pandemic which swept through country after country, Elon Musk was focused on maintaining a broader perspective. After mathematical models predicted that 40 million people alone would die in the United States, prolonged lockdowns, mask mandates, and the race to create a vaccine ensued. During this time, Elon took to Twitter to share his thoughts on the situation.

At first, when responding to people who were worried about the pandemic, Elon claimed that the panic surrounding the flu was 'dumb.' Soon after, he predicted that the cases would be close to zero, even though reported death rates continued to climb. In another tweet, Elon claimed that children were essentially immune to the virus, pointing out that the majority of children barely got sick from the virus.

As the COVID-19 crisis continued on, Elon constantly voiced his opinion against the lockdowns and refused to close his offices—until he was forced to do so. Elon kept claiming that people were being placed under house arrest, and that people should be free to move on with their lives as long as the necessary precautions were taken. Holding a position of support towards vaccines in general, the mandating of a vaccine was what Elon had issues with. Later on, when state governors defied the federal government's demands for continued lockdown and insisted that the pandemic generated more alarm than necessary, Elon publicly celebrated what he saw as a necessary return to normalcy.

Misrepresenting Elon through his tweets is easy. Many of his opinions shared are often part of his mental process and do not entirely represent his final thoughts or opinions on a subject. During his interview with Joe Rogan, however, Elon was able to explain his stance a little bit more. He explained that he saw COVID-19 play out in China before it hit North America and recognized the true state of affairs before everyone else in America. As a result, his predictions about mortality rate were lower than the original predictions made by mathematical, medical models and predictions. Although he agreed that COVID-19 would adversely affect the elderly and sick, Elon believed that the entire situation became all about the panic, which led to ongoing policies that wouldn't be easily reversed.

In the same podcast interview and in various tweets, Elon shared his concerns about the byproducts of the lockdowns—from job insecurity, economic instability, mental health issues, and delayed 'elective' surgeries. Instead of sweeping measures, Elon preferred to promote proper hygiene, masks, self-care, and personal health measures such as remaining at home when sick, citing Japan as a good model for best health practices.

In short, Elon believed that COVID-19 offered the world a chance to learn how to deal with the inevitable more serious illnesses that may arise in the future. His tweets were a part of his process of managing the conflicting information coming in about the pandemic, and a way of parsing out his thoughts on the solutions that had been enacted. True to his character, Elon instead focused on what he could do practically for the world. Despite his controversial opinions on the COVID pandemic, Elon provided New York city with ventilators to help the hospitals, which provided support for less critical patients. He even redirected the production of his factories to develop and produce more FDA-approved ventilators. Although the FDA ventilators were never conveyed fully through production, the prototype, which was presented on internet channels, looked very promising. When the worst of the pandemic was over, the former mayor of New York City, Bill de Blasio, thanked Elon Musk for his contribution to their medical system.

A Dangerous Twitter Poll

Since Elon uses his social media as a platform to voice his opinions in a democratic space, he sometimes finds himself in trouble—not only with the public, but with his board members and the authorities, as well. The billionaire tweeted in 2018 that he wanted to take Tesla private again, paying $420 a share. The tweet raised issues between Elon and the company's board, which led to a lawsuit that eventually was settled in court. As a result, Elon lost his seat on the board. After this 'announcement' on social media, the court determined that all of Elon's posts related to the company would be required to be vetted by his lawyers.

Due to the nature of this tweet, the Securities Exchange Commission (SEC) filed a lawsuit against the billionaire for civil securities fraud. On top of that, the shareholders filed a class-action suit claiming a loss of capital. For the SEC complaint, the court sentenced Elon to pay an indemnization of its shareholders for the loss of value of their stock, and the final deal was set and revised in 2019. However, in 2022, Elon tried to end the agreement, which required both him and Tesla to pay $20 million each.

As for the lawsuit from the shareholders, it is still ongoing, something that Elon Musk is not shy to talk about. Because of this, the accusers want the court to forbid the Tesla CEO from commenting on the case until a settlement has been reached. At a TED conference in 2022, he claimed that the SEC knew that he had funding secured and, even so, filed the lawsuit against him in a very unfair manner. According to reports and interviews given by Elon, a Saudi Arabian fund had been initially secured, but the SEC still would not process the claim. None of this information has been confirmed by any sources, though the billionaire claims that banks forced him to take the deal or they would forfeit any funding. If he had not taken the deal, Tesla would have gone bankrupt since it was under financial pressure at the time.

The 2018 drop in company stock price thanks to a tweet would be repeated in 2020. Once again, Elon had to deal with the SEC because of a post on social media. On May 1, 2020, Musk tweeted a poll in which he asked his followers if he should sell Tesla shares. The SEC claimed that this was a violation of their deal regarding publicly speaking about the company on social media, and questioned if the information was authorized by his lawyers.

Additionally, Elon has been accused of insider trading with his brother Kimbal, due to the same controversial tweet. Currently under investigation, there has been some speculation that Kimbal knew about Elon's poll tweet beforehand. Just one day before Elon's tweet was posted, his brother—who is also a board representative—sold $108 million in shares. As expected, after the tweet poll, the stock value plummeted. In defense, Elon claims that the SEC is harassing him and preventing him from exercising his freedom of speech. He declares that he will continue to fight the lawsuits and that he will put an end to the entire ordeal, even though he did not start it.

Bill Gates Versus Elon Musk

One person with whom Elon Musk is continuously sparring on social media is Bill Gates, the former richest man in the world and owner of Microsoft. Considered to be a major philanthropist and entrepreneur, Bill Gates (no doubt) recognizes within Elon familiar skills and qualities. Still, the two men have clashed over many topics. It all began when Gates criticized the price range of Tesla cars in an interview.

During the COVID-19 pandemic, the two scuffled online over the number of deaths and the efficiency of vaccines. While Gates is a defender of the vaccine and largely contributes to science through social and political action, Elon criticized the actions, strategies, and procedures taken by authorities when the vaccines were made available. Regarding Elon's comments, Gates claimed that Elon had no idea what he was talking about regarding vaccines and that he should stick to what he did best, which was making cars.

More recently, Elon has claimed that Bill Gates has bet a lot of money, over $500 million, that Tesla's stock value would fall, which is called *short-selling* or *shorting*. When Elon asserted his belief that Gates was working against him, Bill Gates agreed. The Microsoft founder, however, offered Elon an opportunity to discuss a variety of options for various philanthropic causes. Elon, in response, claimed that he could not take Gates' request for philanthropy on climate change seriously if he was betting against electric cars. In response, Gates said that there are other ways to diversify his options. Gates asserted his belief that, with the increase of competition amongst electric vehicles in the market, the prices will decrease, eventually leading to a more sustainable industry.

While Elon wouldn't disagree with that, it is clear that Elon hoped Gates would support American enterprises, such as his own.

Regarding Elon's latest potential acquisition, Twitter, Bill Gates has gone on record claiming that the billionaire might actually be doing more damage than good. Gates says that Elon has been brilliant in setting his companies, but he wonders whether Elon will be equally successful running Twitter.

Taxation Troubles

Another topic that Elon is open to discuss at length, on Twitter, is taxation. The billionaire has constantly raised concerns in regard to taxes, claiming California to have an abusive taxation system that penalizes entrepreneurs. Elon eventually moved Tesla's headquarters from California to Texas to avoid such hefty levies. Despite the move, Elon still had strong sentiments about how people viewed the wealthy and taxation in California.

For instance, in 2021, a Twitter feud sprang up between Elon and Senator Elizabeth Warren. Elon claimed that he would pay the highest value in taxes any American has ever paid. Warren, who has been an advocate for taxing large fortunes, presented Elon with some pushback, tweeting that the taxation system was flawed. One of her replies stated, "Let's change the rigged taxed code so that 'The Person of the Year' will actually pay taxes and stop freeloading off everyone else". After everything Elon Musk has achieved, his philanthropic donations aside, the implication that Elon is a freeloader led to further divisive tweets.

Later, when asked about the federal government's plan to pass a bill for which he would have to pay over $50 billion in taxes, Elon warned his viewers, readers, and middle-class investors that "they eventually run out of people's money and then go after you". Elon, who does not receive a salary and only owns stock in his companies, found support from both sides of the political aisle, as various politicians also questioned the legality and potential implementation of taxing capital gains as yet to be realized. Still, Elon's skepticism about taxation policies has continued to cause debate on social media and mainstream media channels.

Other Controversies

As a man who is active on social media, these previous examples are not the only controversies in which Elon Musk was involved. With his awareness of ongoing cultural hot topics, Elon always has something to say about current affairs, unafraid to express his opinion. Elon offended Senator Bernie Sanders when he said that he "keeps forgetting that he was still alive." More backlash arose when Elon said that neutral vocabulary sucks, a comment that drew critique even from his then partner, Grimes.

Other polemic tweets made by the South African-born entrepreneur include memes making fun of people, discussing birth rates, cryptocurrency memes, and insulting public transportation, which led to several complaints from his followers. Elon has also created havoc in specific markets, such as when he claimed he would buy Coca-Cola to "put the cocaine back in." The billionaire also came under critique by a certain sector of the public when he appeared in the Joe Rogan podcast and smoked marijuana. Elon came back with a quip, as you might expect: "As anybody who watched that podcast could tell, I have no idea how to smoke pot or anything. I don't know how to smoke anything, honestly".

Given Elon's interest in meme culture and speaking his truth, when it comes to politics, some may find Elon's usually tight-lipped stance surprising. When his opinion was asked on other political matters, he generally issues a short statement, but says he would rather stay out of politics. However, Elon has admitted that he donates to both the Republican and the Democratic parties, saying that money needs to be involved in order to have his voice heard. Even though he was part of President Trump's advisory council in 2017, he resigned due to the former president's actions regarding climate change.

International affairs also seem to be something that cannot avoid the billionaire's commentary. To the delight of his fans, Elon has been positively vocal about the trucker's protest in Canada, declaring that Canadian trucker's *rule* and they conveyed *strong commitment* as they protested in the dead of Canadian winter. Of course, Elon has also issued opinions on the Bolivian invasion and Russia's invasion of Ukraine, highlighting ongoing global conflicts for his fans. All in all, it seems that Elon is not stopping with his tweets anytime soon.

Television and Film Cameos

Thankfully, Twitter isn't the only means that Elon enjoys communicating. As a public personality, it is common for television shows and movies to request for Elon's presence, when he is available. Elon shared that he enjoys joking around with self-deprecating, dry humor. He generally performs the cameos in television representing himself—and will make fun of himself, as well. Shows such as *The Big Bang Theory*, *The Simpsons*, and *Young Sheldon*, among others, have all starred Elon Musk as himself. One of Elon's most hilarious debuts includes his appearance in the 26th season of *The Simpsons*. In an episode entitled "The Musk Who Fell to the Earth," Elon and his inventions are displayed in a humorous fashion. It hints at a future *Hyperloop* (which would later be built by the Boring Company), makes fun of his electric cars, and jokes about his rockets.

Later on in 2022, Elon also attempted his first opening speech for *Saturday Night Live* on Mother's Day, where he presented his mother to the public and joked about his social awkwardness. With stilted speech and a small smile on his face, Elon quipped about his Asperger's and almost robotic existence, stating, "So I'll make a lot of eye contact with the cast tonight. But, already, I'm pretty good at running human in emulation mode".

Overall, Elon's few brushes with television and film have proven him to be an easy-going, humble man who isn't afraid to laugh at himself, and embracing his awkwardness. When taking his personal life and dreams into account, it is clear that Elon cannot be summed up by his tweets alone. While his presence on Twitter might sometimes cause controversy or stock price declines, Elon is trying to promote a democratic space where everyone can explore possibilities together and share their voices.

Chapter 7:

The Elon of Today

I'm interested in things that change the world or that affect the future and wondrous, new technology where you see it, and you're like, wow, how did that even happen? How is that possible?

–Elon Musk

As you may imagine, a man as busy as Elon always entertains new ideas, projects, and aspirations. According to his friends, family, and coworkers, his mind never stops ticking—there is always an innovative idea that sparks new pathways. The latest plot in Elon's timeline was the potential acquisition of Twitter. Elon's announcement generated a lot of attention. However, although Twitter is the latest trend on Elon Musk's agenda, there are several other projects which he is currently pursuing.

From his ongoing work with The Boring Company to further development on Tesla and the SpaceX rockets, Elon finds himself occupied. Admitting that he usually works a minimum 60 hours a week, Elon is proud of his hard work and the progress that is showing from his varied projects. The Boring Company, for example, has taken on new projects, hoping to build train and car tunnels to improve mobility around the country, and even the world. Elon is also in the midst of launching a new car model for Tesla, a vehicle that already has half a million pre-orders.

Life isn't all sunshine and daisies for Elon, though. Musk is still dealing with an ongoing SEC investigation due to claims that his brother, Kimbal, benefitted from the dip in Tesla stock sales. There is also ongoing experimentation at Tesla that has caused some setbacks. While Elon might envision a future where 100% autonomous vehicles exist, reducing vehicular accidents due to reliable computing, tests have been going on for years with not as much success as the innovator hoped. Still, if we have learned anything about Elon, it is that the man is stubborn and willing to learn from his mistakes. As a result, he stands firm on his vision for the future and believes that he will be able to achieve his goals.

The Boring Company

As Elon continues to juggle the numerous companies he owns, each one focuses on their own individual endeavors. Currently, The Boring Company has two active construction operations, and several more under discussion. The first venture, which is estimated to start in the second quarter of 2022, will create a tunnel beneath the Resorts World Connector. This tunnel will provide underground access between the Resorts World and the Las Vegas Strip, hopefully minimizing traffic congestion on the ground by providing alternative transportation access underground. According to the company's website, the travel time will hopefully take somewhere between one to four minutes, depending on the origin and destination of the traveler.

The second project, which is already underway, is also in Las Vegas. It is called the *Vegas Loop*, and its main objective will be to connect the main points of access and attractions. For the ever-busy city, the airport, stadium, and downtown Vegas are commonly connected points of travel, resulting in traffic jams. Thanks to The Boring Company's tunnel, travel time will be substantially reduced. The project estimates that there will be a reduction of more than 25 minutes between these three focal zones of the city, which will help relieve the congestion in the area drastically.

Other developments for The Boring Company include building loop tunnels in several other cities. Although these loop tunnels will mainly service Tesla, electric vehicles will be able to traverse at high speed through these tunnels, relieving the above ground roads of some traffic. Elon has presented his projects to cities such as Miami and Fort Lauderdale in Florida, as well as other metropolises and states such as Chicago, Baltimore, and Texas.

Neuralink

Neuralink remains, for the most part, a piece of futuristic hardware as yet to be commercially produced. With the technology still under development, the company's main goals, currently, are to create a neural implant that will help treat brain disorders and other health defects, and eventually building towards a symbiosis with AI. This requires a high degree of understanding in the biomechanical and engineering field, and remains a challenge for Elon's company, as well.

The implant, called the Link, is said to be too fragile to be held by a human hand; so, Neuralink has also created a specific robot to insert the threads required by the Link in the brain. Called Brain Computer Interface (BCI), this prototyped microchip would operate with electrical communication similar to that of the brain, which would allow it to operate, theoretically, without concern. According to Neuralink, after a two year delay, it finally hopes to start the first human tests in 2022. Whether we will find ourselves linked to a matrix within our lifetimes remains to be seen, but despite the high rate of failure in this field, Neuralink is forging ahead with experimentation and iteration of the Link.

Tesla

As mentioned prior, part of traveling safely in a tunnel requires shifting to electric-powered vehicles that do not admit gasses or fumes. Due to rising awareness about climate change and the importance of lowering our carbon footprint, alternative approaches to transportation are considered and developed every day. Tesla continues to work towards a sustainable future, as Elon works to make Tesla the ultimate brand in electric cars.

As a result, production at the Tesla factories continues at full speed. The car company is currently commercializing four car models: Model S, Model X, Model 3, and Model Y, which are available for online orders. They are also developing other models, including a truck and a second generation of the Roadster model.

In order to support rising demand, Elon's time has been taken up with managing the production side of Tesla. Thanks to Elon's hard work ethic, the company is also expanding its factory sites. After opening

Gigafactory 3 in Shanghai, China, Tesla has announced that they will be opening another sister plant next door to the Chinese factory at some point in the future. Meanwhile, Tesla opened another factory in Germany on March 22, 2022, called Gigafactory 4, as well as Gigafactory 5 in Austin, Texas, in April of the same year. The factory in Texas is currently the company's headquarters, which was transferred from California due to Elon's claim that California's current tax laws punish private business owners. California's loss is Texas's and Germany's gain. Both of the new factories will produce cars and lithium-ion batteries in the future, as part of Elon Musk's project to increase participation in the sustainable market.

Aside from building new factories, the company is also focused on growing Tesla's brand awareness in the market by increasing the number of stores and convenience products related to the brand. Today, they sell a variety of Tesla products on their website exclusively to the United States, such as tequila, model cars, and other brand-related memorabilia.

Other plans for Tesla include signing a deal with the Brazilian mining giant Companhia Vale do Rio Doce to supply nickel: an essential element for fabricating the company's cars. This strategy is based on the increasing demand for the product that researchers expect to be short in availability by 2026—even though the demand is set to increase up to 40% by 2040. Tesla has been signing contracts with mining companies for the past few years, anticipating the shortfall of nickel. One of the advantages of signing with the Brazilian company is due to the fact that Companhia Vale do Rio Doce mines nickel in Canada, making supply easier to transport to the factories in the United States.

However, the contract comes with a major caveat: the company needs to extract the mineral sustainably, as published by Elon Musk in a Tweet in 2020. Although the details of the negotiation are unknown, the millionaire promised a long-lasting and 'gigantic' contract to any company that was able to extract the product efficiently and in an environmentally safe way. Since Elon is committed to long-term sustainability, his projects usually involve him taking into consideration environmentally friendly practices.

Beyond simply making cars, however, Elon hopes that Tesla, Inc. will become an energy solution company for several markets, especially when it comes to storing solar, wind, or other forms of alternative energy. Lithium-ion batteries might be our future, but Elon and Tesla believe that they can be developed further. This will involve revisiting the quality and technology behind high-powered batteries so that they will not only be equipped to store more energy, but also recharge faster. Overall, it looks like Tesla is going to be extremely busy for the foreseeable future.

SpaceX

Despite his concern about planet Earth and the future generations of humans living on it, Elon is still dedicated to the idea of exploring and colonizing space. As a result, SpaceX continues to play a large role in Elon's plans, moving forward. By 2022, SpaceX was able to launch more than 2,500 Starlink satellites. Although, Elon hopes to launch countless more in order to reach his goal of providing worldwide coverage of Starlink's services.

This goal might take longer than expected however, especially as other worldly situations arise. For example, when Russia invaded Ukraine in early 2022, the billionaire committed resources to Ukrainian citizens. In particular, Elon facilitated the company's satellites to provide free internet for the Ukrainians after receiving a request on Twitter from Mykhailo Fedorov, one of Ukraine's ministers. Elon's charitable gift has enabled Ukrainians to stay in touch, despite Russian hacking attacks on their regular internet services.

Beyond offering Ukraine assistance, Elon has continued to carry forward NASA's work. In May 2022, Elon brought seven astronauts back from the space station with his Space Dragon shuttle. He has reportedly been encouraging NASA and American authorities to think bigger when considering space exploration. With time and support from America and other country's space agencies, Elon's dream to reach Mars might be achievable within this generation.

In the meantime, SpaceX focuses on a slew of missions planned for 2022—rideshare flights, resupply missions to the International Space Station, and observation satellite launches. SpaceX will also be carrying different groups of astronauts to the ISS over the year as well as field a rocket for an ice mining mission on the south pole of the Moon.

With all of these important events, you might think that SpaceX has its hands full. Still, it is making sure to focus on other important parts of the business, such as raising brand awareness. The SpaceX website offers live updates and GPS feeds for all of its launches, as well as video feeds of important lift-offs. On the official SpaceX online store, you can purchase space-related memorabilia, such as t-shirts, kid's spacesuits, backpacks, mugs, and hoodies with the company's motto and designs. One T-shirt in particular, with its logo of "Occupy Mars," can't help but remind you of Elon's tongue-in-cheek humor.

Twitter

Elon's newest and most inflammatory business move was to announce that he planned to acquire Twitter, a public company. Despite the fact that multiple social media websites have previously been privately owned, Elon's announcement on April 14, 2022 was met with equal amounts of shock and joy. Since January of 2022, he had been buying up the company's shares. With a total of 9.1% in shares, the billionaire became the platform's largest stakeholder.

Concerned about a potential hostile takeover, the company's board of directors then invited him to be a chairman, which he initially accepted. However, he would not be able to hold more than 14.9% of the company's shares while Twitter remained a public company. Furthermore, as chairman of Twitter, Elon's hands would be tied by bureaucratic red tape, placing more obstacles between himself and his goals—to make Twitter a democratic, open-source space, free from bots and an opaque algorithm.

Thanks to his foresight, Elon refused to take the bait, backing out of the offer. Instead, he decided to take the company private again by buying its shares for approximately $44 billion. If the board refused, Elon would have to consider investing in alternative social media websites. In fear of Elon potentially offloading all of his shares on the market, and causing the Twitter value to crash the board considered the acquisition. Recognizing their position and their responsibility to the shareholders, the board accepted Elon Musk's proposal to become its owner on April 25.

By May 2022, the entrepreneur secured loans to procure the social media website with the intention of privatizing it again. Using loans from banks, gaining partners, and drawing on his own money from Tesla shares, Elon was able to raise the funds required for the deal. With private ownership of Twitter, Elon believed that he would reinforce the right to free speech.

Claiming that his top motivation for Twitter's acquisition is free speech, Elon believes that Twitter has created an unhealthy algorithm that does not allow for unbiased, free-flowing information. In the wake of the news, previously banned individuals found themselves welcomed back to the platform. Others noticed that their follower counts were rising or decreasing, giving rise to rumors that Twitter was finally cleaning up some of its botting and shadow-banning issues. While many people celebrated the return of previously shunned people to the platform, others voiced their concerns that this might encourage misinformation and jeopardize the movements that the establishment had hoped to encourage, such as climate change or pro-vaccine policies.

Elon, however, isn't afraid to tackle the difficult problem of balancing democratic speech and ensuring that information is properly shared. Acquiring Twitter has brought back a set of abilities that Elon used back in the day while navigating and dealing with internet companies. While he hasn't done much work with programming internet solutions since he sold PayPal in his early days, Elon feels that he is up to the task of reinventing Twitter. Unfortunately, the deal would never eventuate as Elon terminated the merger agreement due to apparent false and misleading misrepresentations that were not brought to light upon entering the deal.

Elon had announced some great changes for the platform, some of which included intentions to open source the social media's algorithm to increase transparency, and charge governments and companies a fee for use of the platform. He had also stated that he does not trust the company's present management, which likely meant they would have been released from their duties as soon as the takeover was finalized. Finally, the billionaire has said in several interviews that he believes that Twitter has great potential that just needs to be unlocked. To that end, Elon had convincing plans on removing automated bots, enhancing the platform with new features, and authenticating all humans who want to use the platform.

Along with releasing the company's algorithm to the public, Elon also motioned to create an 'edit' button to enable posts to be altered and make sure that the free speech he advocates for is in accordance with the country's laws. This means that the platform's users would be susceptible to local legislation when evaluating their Tweets, which could mean that misinformation or hate speech would be moderated legally.

In a swift turn of events, the acquisition rumors finally came to a halt in late October with Elon revealing that he had indeed acquired Twitter. An announcement from Twitter via email uncovered that the company was letting go of a good chunk of employees including some of Twitter's top executives such as the former CEO Parag Agrawal, chief financial officer Ned Segal and chief legal officer Vijaya Gadde. Since taking over, Elon has attempted to launch his subscription service, Twitter Blue, allowing users access to verification alongside a monthly $8 fee. The service has since been put on hold due to an influx of imposters and fake accounts claiming to be famous celebrities, companies, athletes, and even Tesla.

It will remain a mystery as to how all of these issues will be achieved since moderating speech remains a knotty problem for many social media sites. Still, if there is anyone who can achieve the supposedly unachievable, it's Elon Musk!

Chapter 8:

To the Moon and Beyond

We're going to make it happen. As God is my bloody witness, I'm hell-bent on making it work.

–Elon Musk

While it is impossible to talk about Elon without mentioning his current activities, it is also unthinkable to mention him without discussing what he plans for the future. Whether they are just ideas that he tweets about, such as creating a university, or ideas he has already set in motion, such as sending man to Mars, Elon Musk seems to always be looking ahead of his time. With his gaze fixed firmly on the horizon, Elon is always thinking about what might be the next big problem to solve, whether it involves space travel, autonomous cars, or artificial intelligence.

Colonizing Space

Maybe you have thought about what the red planet is like—or maybe not. The fact is that Elon Musk has his eyes set on sending astronauts to explore Mars, hopefully by 2025. He believes that humanity will one day become a multi-planetary species. Unfortunately, travel to Mars is difficult, and that's not even including the complications of nurturing life on the red planet. It would take around six months of close-quarter travel to get to Mars, resulting in a lot of complications that need to be overcome: from producing air for the crew to supporting mental health for the long period of travel. Of course, Elon has been considering these issues for quite some time, so he has been working on resolving some of the largest hurdles to long-term space travel. The billionaire thinks that it is possible—and his ideas are slowly coming to fruition.

According to the SpaceX webpage, colonizing Mars is one of the company's main objectives, although not an immediate one. They expect to send a crew to the planet with the Starship rocket, which SpaceX estimates will be the most powerful rocket of its time when iterations of the ship are concluded. The mission includes taking equipment to Mars and using the planet's available water and carbon dioxide to fuel the ship before returning back to Earth. Even if this does not eventuate as planned, an alternative strategy for refueling has been considered, with fuel tankers strategically located within orbit, if necessary.

When will this historic space voyage happen? That is still up in the air. While SpaceX executives believe that the first crew could be on Mars as early as 2030, NASA argues that the timing is too optimistic and probably won't be possible until later, around 2040. While the plan to colonize Mars remains uncertain, both SpaceX and NASA hope to put man on the moon in the not-too-distant future. NASA estimates that man can be taken to the moon's south pole as early as 2025. SpaceX executives agree, noting that the moon would have a crew landing on its surface before the Mars landing. However, both agencies have not yet determined a date for a moon landing.

Working together, both SpaceX and NASA will be able to speed up the process and divide the cost of putting humans on the moon. With NASA's expertise and Elon's resources and technology, space exploration could be closer than ever. When asked about Mars, Elon once said, "I think it would be great to be born on Earth and to die on Mars. Just hopefully not at the point of impact". *Never change, Elon.*

Flying Electric Airplanes

Given Elon's adoration of transforming traditional notions about travel, it is no surprise that the innovator has mentioned electric airplanes once or twice. Elon's eye for innovation could transform how airplanes operate and perform. For starters, airplanes rely on fossil fuels, contributing to ongoing air pollution. Furthermore, while some specialized jets and airplanes can reach incredible speeds, commercial flights with supersonic jets are rare.

Although electric airplanes have no doubt been in discussion for a long time, the public was made aware of the possibility in Iron Man 2, where Elon had a cameo involving a short discussion with Tony Stark about an option for an electric airplane. Since then, Elon has been linked to the development of electric airplanes. In 2021, Elon admitted that the notion of electric airplanes had occurred to him, tweeting, "I'm so dying to do a supersonic VTOL electric jet! But I already have way too much on my plate. Any more work and my brain will explode".

Despite his repeated statements that he is too busy to give electric airplanes proper consideration, Elon has shared his ideas on a variety of occasions. In a conference and various interviews, Elon elaborated on what his ideal electric airplane would look like. It would be an electric supersonic jet with cool, minimalist details. Although he admitted that he has a design in mind, Elon refused to give any more details. What he has suggested is that his electric airplane would utilize vertical take-off and landing (VTOL) technology and would be powered by a highly advanced battery system. The speeds of the jet would reach around 768 miles per hour. To put this into perspective, the average airplane currently achieves 575 miles per hour max.

Unfortunately, to achieve lift off, maintain propulsion, and zip along at supersonic speed, the airplane will require a highly dense battery. Elon went into the details of the problem, stating that the current Tesla Model 3 has a battery density of 250 watts per hour per kilogram. For the plane, they would require 400 watts. Still, Elon is hopeful that development on battery life and operation will increase in sophistication over the next few years. Until then, Elon has dedicated himself to his current projects. "It would be a fun problem to work on at some point," he said, "but we've got a lot to do over the next few years, so we've got to focus on these things. Get them right, and then maybe one day do that". For now, we will have to wait for continued improvements in battery density, and for Elon to finish his current assignments. One day though, we might experience our first electric supersonic jet flight.

Driving Autonomous Cars

As mentioned earlier, another one of Elon's dreams includes having self-driven cars with little to no interference from the driver. Tesla claims that all of its vehicles have the hardware to be autonomous, but the software is not yet available because they have not found a reliable option for autonomous driving. Even so, the company has made huge progress in developing sophisticated autopilot features, allowing cars to be somewhat autonomous under the supervision of a driver. The company's vision is that a person will be able to get inside their car and simply tell their Tesla car where they want to go, and the vehicle will take them using optimized routes. If the driver does not say anything, the car will access the individual's schedule and take them to exactly where they need to be.

Even though this might seem like far-fetched science fiction, Tesla claims that these goals are quite achievable. For starters, Tesla expects to deliver a feature called "auto parking", in which the driver will exit their vehicle, and the vehicle will enter "parking mode." In this mode, it will search for an available place to park. How the driver will find it afterward is still a mystery, but more than likely the car's final parking spot will be relayed via GPS signal to the driver's mobile device.

While the idea of autonomous driving or parking sounds exciting, there are many hurdles for Tesla to overcome. Although there have been many advancements in the field of autonomous driving, the software is still under development. Several accidents, including crashes, have been registered during testing. As a result, Tesla has been unable to release their proprietary software to the masses. In turn, it is unclear when it will be safe and made available. While the autonomous car has impressed testers, the ongoing software issues and limitations still require further advancements. There seems to be the need for plenty of intervention from the driver with current specifications, but the company is currently hiring test drivers worldwide to put the cars' features through their paces.

Creating Ethical AI

Due to Elon's trepidations about the impact of AI on society and technology, Musk has begun an ethical AI project with other experts in the field. The humanoid robot called Optimus will be designed to take on unwanted tedious and repetitive jobs. According to Elon, the Optimus robots will prove more popular than his Tesla cars. The project was presented in August 2021, and is estimated to be available to the public by 2023. As usual, many critics have voiced some skepticism about how ready the market is for high-functioning industrial robots, even though many robots and machines are already in use currently.

According to Musk, however, plans for security measures have been implemented in order to ensure safety, particularly in terms of AI operation and containment. Management of the Optimus robot will have access to deactivate the AI, thanks to software security features. Elon, who has already shared deep concerns and warnings about the dangers of developing AI, made certain to include methods to keep the Optimus robot limited. The business mogul believes that, if not treated carefully, AI software may pose a greater threat than climate change, out-thinking and overpowering humans. Thus, Optimus robot will be deployed with ethical AI software, designed with inherent limitations intended to protect humanity.

Although this project seems promising, information regarding Elon's ethical AI project remains sparse. Other than a speech he gave in recent interviews on the dangers of AI and his investment in ethical AI, Elon has not shared details on the Optimus robot, nor has a model been presented to the public. Initially, Elon believes that the robot may prove to be expensive due to lower demand at the beginning of production.

Conclusion

There's a tremendous bias against taking risks. Everyone is trying to optimize their ass-covering.

–Elon Musk

Whether you are a fan of Elon or not, there is no denying that his ambitions and accomplishments are impressive. He has managed to build, develop, and trade incredibly successful businesses. He has built an empire and amassed billions of dollars in fortune. He has continued to push the envelope of scientific, engineering, and technological development... and this is still not the end of it!

The 50-year-old has not only succeeded as a wealthy businessman, but has become an icon of innovation and free-thinking. Individuals are constantly checking Elon's Twitter to view his tweets, waiting for his opinion on current issues. Both Elon's admirers and haters lay in wait to see what Elon is going to do next, whether it be in regards to political matters or company developments.

As witnessed throughout his life, Elon Musk has proven himself a determined man who goes after what he desires. He overcame childhood difficulties in school and a harsh upbringing at home to become one of the most influential men on the planet. Thanks to support from his mother and siblings, Elon has been able to pursue his goals while also pushing his family to achieve their own dreams. With a large family of seven children, which he cherishes, Elon has continued to invest in the future generation by providing educational alternatives that will unlock many capabilities for the students.

As CEO of Tesla and SpaceX, Elon has his plate full, and he still manages to support with the management of The Boring Company and Neuralink. It is uncanny to realize that somehow Elon succeeds at achieving so many tasks. With a mind travelling a million miles per hour, Elon shows how much a person can achieve if they simply open their world up to possibilities. Besides working on new ideas and projects, Elon still finds time to be with his family, participate in television and movie cameos, play video games, and read. In between all of this action, Elon still maintains a highly active presence on Twitter, personally answering questions, replying to memes, and sharing his opinions on different matters.

Elon Musk is like a perpetual motion machine. It is a wonder that he even finds the time to sleep. Juggling what feels like a million and one projects, Elon has his hands full overseeing his myriad of companies—Tesla, The Boring Company, SpaceX, Neuralink, and Twitter—from developing energy efficient and sustainable electric vehicles to encouraging alternative energy production to boring tunnels for improved transportation. With the Starlink operation now finalized and Neuralink developing technology for brain chips, Elon is committed to a better and healthier world sustained by ethical technological practices. All of this is led by Elon Musk and his visionary ideas. *How does he do it?* That's the real mystery.

However, it seems that, although Musk keeps on trying to improve the world and make it a better place to live in, people keep remembering his controversies and doubting his intentions. While it is easy to focus on the lawsuits, polemic tweets, and internet battles with other celebrities, it is crucial to remember that Elon is a man who, like everyone else, is prone to failure, despite good intentions. As someone who has apologized, admitted failure, and learned from mistakes, Elon has proven himself a model for visionaries and dreamers. Furthermore, Elon's passion for truth has driven him to pursue equality in the internet space, even if (sometimes) he might go a little overboard.

How Elon will be remembered in the future is still unknown. Time will tell how much his philanthropic efforts and technological innovations will impact history overall. Still, even with his current achievements, Elon will have made his mark on history. Whatever happens in the years to come, Elon remains cautiously optimistic. Instead of preaching a message of doom and gloom, Elon intends to work toward the future he wants to see. As he once said, "I want to be clear: I'm not trying to be anyone's savior. I'm just trying to think about the future and not be sad".

www.ingramcontent.com/pod-product-compliance
Lightning Source LLC
Chambersburg PA
CBHW041309110526
44590CB00028B/4302